Doubletree Hotel Fire
New Orleans, Louisiana

Investigated by: Jeffrey M. Shapiro, P.E.

This is Report 008 of the Major Fires Investigation Project conducted by TriData Corporation under contract EMW-86-C-2277 to the United States Fire Administration, Federal Emergency Management Agency.

 **Homeland Security**

Department of Homeland Security
United States Fire Administration
National Fire Data Center

U.S. Fire Administration Fire Investigations Program

The U.S. Fire Administration develops reports on selected major fires throughout the country. The fires usually involve multiple deaths or a large loss of property. But the primary criterion for deciding to do a report is whether it will result in significant "lessons learned." In some cases these lessons bring to light new knowledge about fire--the effect of building construction or contents, human behavior in fire, etc. In other cases, the lessons are not new but are serious enough to highlight once again, with yet another fire tragedy report. In some cases, special reports are developed to discuss events, drills, or new technologies which are of interest to the fire service.

The reports are sent to fire magazines and are distributed at National and Regional fire meetings. The International Association of Fire Chiefs assists the USFA in disseminating the findings throughout the fire service. On a continuing basis the reports are available on request from the USFA; announcements of their availability are published widely in fire journals and newsletters.

This body of work provides detailed information on the nature of the fire problem for policymakers who must decide on allocations of resources between fire and other pressing problems, and within the fire service to improve codes and code enforcement, training, public fire education, building technology, and other related areas.

The Fire Administration, which has no regulatory authority, sends an experienced fire investigator into a community after a major incident only after having conferred with the local fire authorities to insure that the assistance and presence of the USFA would be supportive and would in no way interfere with any review of the incident they are themselves conducting. The intent is not to arrive during the event or even immediately after, but rather after the dust settles, so that a complete and objective review of all the important aspects of the incident can be made. Local authorities review the USFA's report while it is in draft. The USFA investigator or team is available to local authorities should they wish to request technical assistance for their own investigation.

For additional copies of this report write to the U.S. Fire Administration, 16825 South Seton Avenue, Emmitsburg, Maryland 21727. The report is available on the Administration's Web site at http://www.usfa.dhs.gov/

U.S. Fire Administration

Mission Statement

As an entity of the Department of Homeland Security, the mission of the USFA is to reduce life and economic losses due to fire and related emergencies, through leadership, advocacy, coordination, and support. We serve the Nation independently, in coordination with other Federal agencies, and in partnership with fire protection and emergency service communities. With a commitment to excellence, we provide public education, training, technology, and data initiatives.

 Homeland Security

TABLE OF CONTENTS

SUMMARY
Doubletree Hotel Fire
New Orleans, Louisiana

A six-alarm arson fire on the tenth floor of the Doubletree Hotel in New Orleans, Louisiana, on July 19, killed a security guard, injured ten others, and proved to be yet another example of the need for improved fire safety for high-rise hotels. The U.S. Fire Administration conducted an investigation of this fire as part of its Major Fire Investigation Program TriData Corporation of Arlington, Virginia, performed the investigation.

The fire occurred on a Sunday just after 10:00 p.m. and started in a corridor serving guest rooms on the tenth floor. The floor was unoccupied and undergoing renovation at the time. The cause of the fire was arson. Due to the failure of the automatic fire alarm system, the fire gained significant headway before being detected. The hotel's partial sprinkler system did not cover the area of origin. The delay in detection allowed smoke to spread to at least two other floors before the manual fire alarm was sounded by the guard, who ultimately was killed. Behavior of guests ranged from panic on the smoky floor above the fire to apathy elsewhere due to recent false alarms.

The performance of the building's fire-resistive assemblies and finishes was noteworthy. Fire damage was limited to a single floor; however, smoke damage affected several floors. The fire department effectively used the Incident Command System (ICS) to manage and ultimately control the incident.

Many key issues were identified in this fire pertaining to such areas as fire protection equipment, building construction, fire department operations, and human behavior. A summary of these issues is presented in Table 1. A report is available from the U.S. Fire Administration.

DOUBLETREE HOTEL FIRE
NEW ORLEANS, LOUISIANA
JULY 19, 1987

Local Contacts: Chief William J. McCrossen
Captain George H. Rigamer, Jr.
City of New Orleans Fire Department
317 Decatur Street
New Orleans, Louisiana 70130
(504) 581-6200

OVERVIEW

On July 19, 1987, a six-alarm fire on the unoccupied tenth floor of the Doubletree Hotel in down-town New Orleans resulted in the death of a hotel security guard and the injury of ten others, including hotel guests, hotel staff, and police officers. The fire, which resulted in a total of $175,000 in damage, was caused by arson. Factors contributing to the loss were the failure of the automatic fire alarm system, the lack of automatic sprinkler system protection in the guest room areas, storage in exit corridors, and improper action by an untrained employee. In the shadow of the tragedy, though, resides a success: an estimated additional 150 people were evacuated without injury because of the effective operation of the manual fire alarm system, successfully performance of fire-resistive construction, light fire loading in the exit corridors, and a well organized suppression effort by the City of New Orleans Fire Department.

This report describes and evaluates significant issues pertaining to the fire exclusive of the cause and origin investigation. A summary of key issues is presented in Table 1. The report is divided into five major sections: Overview, Background, The Fire, Analysis of Significant Issues, and Lessons Learned.

This report is not intended to place blame or fix liability upon those individuals or corporations involved in this incident.

BACKGROUND

Construction

The Doubletree Hotel was previously operated as the International Hotel and was constructed in 1973 at 300 Canal Street in downtown New Orleans adjacent to the famous French Quarter (see Figure 1; Figure 2 shows where photos were taken to furnish a frame of reference). The building is a 17-story high-rise and contains 363 guest rooms. The building is constructed of reinforced concrete and appears to qualify as Type 1 construction: non-combustible/fire-resistive. Each typical guest room floor is L-shaped, contains approximately 18,000 square feet, and has 31 guest rooms. The L-shaped corridor connects to three 2-hour fire-resistive stairways, one at each end and one near the corner (See Figure 3).

Corridor walls are made of 5/8 inch gypsum wallboard mounted on 3-5/8 inch metal studs located 24 inches on center. Since the wallboard is not marked to indicate its qualification for use in a rated wall assembly, it is uncertain whether the wall would have qualified as 1-hour fire- resistive construction.

TABLE 1. SUMMARY OF KEY ISSUES

Issue	Comments
Occupancy	High-rise hotel – 17 stories Built in 1973 Approximately 50 percent occupied at the time of the fire
The Fire	Caused by arson Occurred in tenth-floor corridor (unoccupied floor) Killed one and injured ten Ignition involved combustibles stored in corridor Large volume of smoke spread to other floors
Fire Protection Equipment	Automatic alarm failed Manual alarm facilitated evacuation Building was only partially sprinklered (none in area of origin)
Building Construction	Fire-resistive construction helped contain fir Limited combustibility of corridor carpeting and finishes helped inhibit fire spread Ventilation ducts allowed smoke to spread into several floors
Fire Department Operations	ICS effectively used
Human Behavior	"Convergence cluster" occurred – frightened people grouped and took refuge Apathy to fire alarm due to previous false alarms delayed evacuation of some occupants Unwise action by a hotel staff member may have contributed to his death

Corridor doors leading to guest rooms are 1-3/4 inch thick (apparently solid core) wood doors set in steel frames without self-closing devices. Such doors are typically accepted as equivalent to 20-minute fire-resistive assemblies in existing buildings.

At the time of the fire incident, the hotel was undergoing a 5 million dollar renovation, which included mostly cosmetic upgrades and new furnishings.

Codes

The building code in effect at the time of construction was the City of New Orleans Building Code, which was originally written in 1949 and periodically amended. The New Orleans Building Code appears to have been based on an early version of the Uniform Building Code. In addition, the city currently uses the 1985 National Fire Codes, published by the National Fire Protection Association, as a supplement. Other code requirements, according to the New Orleans Fire Marshal, include the 1967 edition of the National Fire Codes, which was adopted by the State of Louisiana as retroactive to all existing buildings within the State. The City of New Orleans has now established a formal program for general enforcement of the retroactive State law.

It is also noteworthy that New Orleans adopted a high-rise provision for new construction in the city's building code in 1975 that required automatic sprinkler protection throughout such buildings. The ordinance did not have any retroactive provisions, and therefore did not apply to the Doubletree Hotel.

Fire Protection Systems

At the time of the fire, the hotel was equipped with a substantial complement of fire protection systems, considering its age. In addition to a partial automatic sprinkler system, the hotel was provided with a combination standpipe for both occupant and fire department use, automatic and manual fire alarm systems, and portable fire extinguishers. Figure 3 shows the location of these systems on a typical floor.

The sprinkler and standpipe systems were fed by a 750-gpm/125-psi fire booster pump located on the third floor. The hotel's sprinkler system protected primarily non-guest areas, including the storage, maids', and janitors' closets on each floor; the lobby; assembly areas; and kitchen areas. Guest room floor corridors and guest rooms were not protected by automatic sprinklers.

The standpipe system consisted of a 6-inch riser in each stairwell with 2-1/2 inch outlets at each floor landing. Three 1-1/2 inch hose cabinets for occupant use were provided on each floor, one immediately outside each stairwell. A 4A-30BC fire extinguisher was provided in each hose cabinet.

Two separate fire alarm systems were installed in the building, one manual and one automatic. The manual system included three pull stations per floor (one at each exit), a buzzer above each pull station, and a control panel in a closet on the third floor. The control panel connected to a remote annunciators, which was located behind the front desk.

According to hotel management, the automatic fire alarm system was installed after the hotel was completed. It included two photoelectric smoke detectors located in the corridor of each guest room floor and combination fixed-temperature/rate-of-rise heat detectors in each guest room and in various other areas. The automatic detectors connected to a control panel that was reported to have been located in the same closet as the manual alarm control panel. The automatic alarm control panel had been removed prior to this investigation. Based on visual observation and the statements of hotel employees, it appeared that the automatic alarm system did not connect to any alarm-indicating devices (e.g., bells, buzzers). No determination could be made what, if anything, would happen if the automatic alarm system detected a fire. This subject will be discussed later in this report. The system wiring consisted of four-conductor telephone cable with solid-core conductors, which does not meet any nationally recognized fire alarm standards.

THE FIRE

On the weekend of July 18-19, the Doubletree Hotel experienced numerous problems with mischievous pranks. During the day and evening of Saturday, July 18, two false alarms were activated on the manual system. On Sunday, the elevators became the pranksters' targets. Several times on Sunday, elevators were left stopped at various floors with the emergency stop alarm sounding. At approximately 10:15 p.m., Sunday, another elevator alarm began to sound.

At the time, approximately 143 of the hotel's 363 rooms were occupied. Included among the guests were a family reunion group with numerous youngsters and a church group consisting of teenagers. In response to the elevator alarm, the building engineer and security guard on-duty were dispatched to find the stopped elevator, each taking a portion of the building. The engineer started at the ninth floor and was to work downward. The guard started at the sixteenth floor and was to work down to the ninth floor. When the engineer arrived at the ninth floor, he found the stalled elevator and silenced the alarm. He then proceeded by elevator to the eleventh floor, where he knew the church group was staying, to look around, apparently thinking that they may have been responsible for the stalled elevator.

When the engineer arrived on the eleventh floor, several of the church group members were in the corridor talking (see Figure 4 for a picture of the corridor). As the engineer moved down the corridor toward Stairway 1, he noticed smoke coming from the ventilator opening adjacent to the stairway door (see Figure 5). Thinking the smoke might be from an air handling unit fire in the shaft, the engineer discharged a fire extinguisher from the adjacent cabinet into the shaft and radioed the PBX operator that he had a fire on the eleventh floor. The operator acknowledged, as did the security guard, who responses, "OK."

Smoke on the eleventh floor quickly became very thick and began to bank down off the ceiling. A church group member reported later that as the smoke continued to build, the group members in the corridor began running to one stairway and then another, being forced back by smoke. Forced to the ground by the descending smoke layer, these occupants began to crawl about, seeking to escape. As panic ensued, a member of the church group assumed command and led approximately 15 of the members into Room 1109, where they attempted to calm each other through prayer. The engineer indicated that he directed some of the occupants to crawl to Stairway 2, then left for the lobby in an elevator to direct the fire department. It did not appear that the fire alarm had activated at this point.

Simultaneously, three guests – a man, a woman, and their teenage son – were in the passenger elevator lobby preparing to leave the ninth floor to check out. As the woman held an elevator and was preparing to put baggage inside, she smelled smoke. She told her husband about the odor, and he told her to get out of the elevator. As she did, she apparently pushed the emergency stop button, sounding the elevator alarm bell. Since they had smelled smoke, the family believed the bell to be a fire alarm, and the man proceeded to the house phone in the elevator lobby to report the alarm and the smoke. The hotel operator advised them to evacuate.

As the man hung up the telephone, the security guard entered the elevator lobby area. Presumably, the guard heard the elevator alarm and was attempting to locate the stalled elevator. The son advised the guard that this was not a false alarm and that he had seen smoke in the elevator. The guard advised them to leave at once, and they exited, apparently via Stairway 2. By this time, a haze of smoke at the end of the long corridor was visible, and smoke was apparently coming from the ventilator opening adjacent to Stairway 3. The son reported that the guard headed in the direction of Stairway 1.

As the family exited, they began to encounter additional occupants in the stairway on floors below the fire floor. Because of this, one can deduce that the fire alarm was probably sounded shortly after they left the ninth floor. On their way down, the family passed two hotel staff members walking up the stairs, whom they told, there really was a fire. Apparently somewhat surprised, the staff members then began to run upstairs. As the family arrived in the lobby, the first due units from the fire department were arriving.

The engineer on-duty had since arrived at the lobby and called the chief engineer for the hotel, who instructed him to shut off the air handling units. The engineer attempted to go up the stairway with the firefighters, but was told to go back. He then returned to the lobby, boarded an elevator, and went to the seventeenth floor to shut off the building's fans. Although the elevator filled with smoke on the way up, he was able to get to the seventeenth floor and access the fan controls. Now trapped by smoke, he called the lobby for help. The chief engineer had arrived and advised him of a means to access a second stairway, which the engineer finally used to escape.

The security guard's actions following his departure from the ninth floor are not known for certain. The most likely sequence of events, as determined by witness statements and physical evidence, is as follows.

In the process of or after departing the ninth floor using Stairway 1, the guard probably doubled back and went into Stairway 2 to access the tenth floor. It is believed that the guard entered the unoccupied tenth floor at Stairway 2 because the manual alarm was only activated at a single location – on the tenth floor immediately adjacent to Stairway (see Figures 6 and 7). Given conditions likely to be present on the tenth floor at this point, it is assumed that the guard would have pulled the alarm as quickly as possible. The timing of the guard initiating the alarm in this sequence is relatively consistent with the other aspects of this scenario, since the family who had exited the ninth floor before the alarm had sounded began to encounter other guests entering the stairwell one or two minutes after leaving the ninth floor. This would have been enough time for the guard to go into Stairway 1, double back to Stairway 2, and get to the tenth floor.

As the guard entered the tenth floor, he probably encountered heavy smoke (considering the volume of smoke that had already spread to the ninth and eleventh floors) but tolerable temperatures. From the pull station at Stairway 2, he was probably able to see the fire through the smoke, which was burning in front of Room 1001.

Fire department investigators speculated that after pulling the fire alarm, the guard may have used one or both of the occupant use fire hoses adjacent to Stairways 1 and 2, which fire investigators found partially removed from their racks but uncharged. However, this investigation indicated that it may have been fire department personnel who removed these hoses during firefighting activity. Fire department investigators also believed that the guard may have entered Room 1029 at some point to seek refuge. Handprints found on the glass were assumed to have been those of the guard perhaps trying to open the window, which in fact was inoperable by design. It is almost certain, that, at some point, the guard actually passed the fire, since he was eventually discovered collapsed and in cardiac arrest on the opposite side of the fire from the pull station at Stairway 2.

The fire is estimated to have originated sometime between 10:00 p.m. and 10:15 p.m., and is believed to have been caused by arson. The fire department determined that there were no other possible ignition sources present and that there was no possibility of an accidental smoldering fire because an eleventh-floor guest had been on the tenth floor to get ice less than ten minutes before the fire alarm sounded and had detected no sign of a fire. Figures 8 and 9 show the area of origin.

At the time of the fire, the tenth floor was unoccupied and undergoing renovation. As part of the renovation process, large wooden cabinets were being provided in each room. The cabinets were packaged in cardboard boxes and were packed with sheets of solid foam. Employees who had been installing the cabinets had stored the empty packaging material, most of which had been flattened and stacked against the wall, in the corridor (see Figure 10). An estimated 10 to 20 boxes that were stacked outside Room 1001 were probably burning when the guard entered the tenth floor.

Fire Department Actions

The fire department received its first call from the hotel operator and dispatched first alarm units at 10:32 p.m. The first companies that arrived on the scene at 10:33 p.m. heard alarm bells ringing and saw some guests evacuating. While en route, they had been advised that a fire was in progress on the ninth floor, a message relayed by the hotel operator to fire dispatch that was probably based on the initial telephone call from the elevator lobby on the ninth floor. Upon entering the building, the first arriving crew ascended toward the ninth floor using Stairway 2. At about the sixth-floor level, they encountered smoke in the stairway and donned breathing apparatus. Continuing upward, they checked each floor along the way. When they entered the ninth floor, there was a haze restricting visibility to approximately 50 feet, but no fire. They then continued to the tenth floor. Smoke became very heavy in the stairway between floors nine and ten.

After checking the tenth floor by cracking the stairway door and determining that there was a working fire in progress, the crew entered the tenth floor with a handline connected to the standpipe. The officer indicated that temperatures at the 3-foot level were only marginally tolerable even with protective clothing. He and his firefighters advanced into the corridor in a crawling position, attempting to locate the area of origin. Some combustion was taking place in the smoke layer over their heads at this time, but they were quickly able to locate and extinguish the isolated fire located in front of Room 1001. The combustion appeared to be mostly smoldering as opposed to open flame.

In the meantime, other arriving units had begun to establish the New Orleans ICS, locating the command post in the security office at the lobby level. By establishing command in this location, the Incident Commander was able to have ready access to resources such as a telephone, floor plans, the hotel manager and engineer, and keys.

Command assigned six sectors to supervise fireground operations: lobby control, ninth-floor sector, tenth-floor sector, reconnaissance sector, rehabilitation sector, and staging. Lobby control was responsible for controlling elevator deployment and maintaining a record of all personnel in the building. As the Incident Commander was advised by the reconnaissance sector of the conditions on various levels of the building, companies were directed to evacuate remaining occupants from floors ten through seventeen, using the ninth floor as the equipment staging area.

Following extinguishment, companies began to require rest breaks, and the rehabilitation sector was used to monitor the status and condition of personnel on break. By maintaining a smooth flow of personnel from staging to lobby control to deployment to rehabilitation, the fireground operation was effective and efficient. The incident was terminated at 03:17 on Monday morning after nearly 5 hours.

Fire damage was contained to the immediate area of origin. In addition, there was heavy smoke damage on the tenth floor, light smoke damage above the tenth floor, and some water and smoke damage on the ninth floor. The total loss was estimated to be $125,000, to the structure and $50,000, to contents, according to the fire department's investigator.

Following the fire, the fire department issued citations to the hotel for illegal storage in an exit corridor and for failure to properly maintain the fire alarm system.

Emergency Medical Services

The New Orleans Health Department provides emergency medical services to the city. Upon arrival at the scene, the first unit established a triage program. The program created a Level 1 staging area in the hotel lobby for patient assessment and care of critical cases. Lower priority patients were referred to the secondary treatment area outside the hotel.

In all, ten patients were treated. The only critical case was the security guard, who was delivered to the staging area in cardiac arrest. Firefighters and EMS crews were unable to revive him. The guard was 35 years old, slightly overweight, and had been found by a firefighting crew in the corridor of the tenth floor. The cause of death was smoke inhalation in combination with a pre-existing cardiovascular condition. Two guests were transported to the hospital suffering from minor smoke inhalation, and a third guest was transported with a burned hand (not directly related to the fire incident). The other six patients, who were treated at the scene, included a hotel employee, two police officers, and three other guests, all of whom had minor injuries.

ANALYSIS OF SIGNIFICANT ISSUES

Fire Protection Systems

Fire Alarm Systems

The manual fire alarm system is credited with giving the early warning that prompted evacuation of occupants from the building. The system was activated at a single location adjacent to Stairway 2 on the tenth floor by the security guard. Once activated, the system sounded buzzers throughout the building and indicated the tenth-floor zone on the annunciator panel behind the front desk.

In contrast, the automatic fire alarm system's apparent failure to detect and alarm probably was the single most significant factor allowing the fire to become a major incident. Based on an analysis of code requirements, it appears that the automatic fire alarm system was not required at the time of construction, nor was it required retroactively; therefore, the system would not have been required to meet any standard. Had the system been properly installed and functional, smoke detectors located in the corridor should have detected the fire long before it became life-threatening. An earlier alarm would have provided additional time for occupants to evacuate before the smoke spread to the upper floors and would have allowed staff the time to investigate and perhaps extinguish the fire before it became severe.

The governing standards dealing with installation of automatic fire alarms in New Orleans are NFPA 72A, "Standard for the Installation, Maintenance, and Use of Local Protective Signaling Systems for Guard's Tour, Fire Alarm, and Supervisory Service." And NFPA 72E, "Standard on Automatic Fire Detectors," published by the National Fire Protection Association. Although the system had been partially dismantled at the time of this investigation, it is still fairly conclusive that the automatic alarm failed to operate. A representative of the fire department – assisted by representatives of a local fire alarm company – attempted to activate the system by blowing smoke into the system's smoke detectors on the day after the fire but was unable to cause an alarm even though the photocell light sources were illuminated. This suggests that smoke detectors probably were not connected to the alarm panel or that the panel had malfunctioned. Furthermore, the automatic alarm system did not appear to connect to any audible alarms, nor did it appear to connect with the manual system, which had audible devices. Therefore, it seems that even if the smoke detectors had been able to detect a fire and the control panel had been functional, the system may only have been capable of indicating a fire at the control panel, which was located in an unoccupied closet on the third floor. Additional significant deficiencies included the use of telephone wire connecting all initiating devices to the control panel, and the placement of all corridor smoke detectors in "dead air" locations in the corridor (within 4 inches of a wall); see Figures 11 and 12 for detector locations.

One other issue although acceptable based on Nationally recognized standards, raises concern regarding the design of smoke detectors. The smoke detectors installed in the Doubletree – Ademco Model 527, listed by Underwriters Laboratories – were a four-wire photoelectric type. The light source for the photocell on this detector was designed to be clearly visible through a large lens located on the bottom face of the detector. According to the building engineer, these lights were always illuminated and were still on when the system was tested the day after the fire. One might well be led to believe that because the detectors were issuing a light from the lens, the alarm system was operational; however, this was not the case.

Typically, a four-wire detection circuit will use two wires for powering detectors and two wires for initiating an alarm. Supervision of the power circuit can be either directly at the alarm panel or at each individual detector. Normally, in this arrangement, a trouble signal will sound at the control panel if power is interrupted or if detectors are disconnected. However, if detectors are powered from a source that is independent of the alarm control panel, the control panel could be disconnected or fail, disabling the trouble signal and alarm-initiating capability, and the detectors could still have power. In such a situation, detectors with a power-indicating lamp would appear functional due to the illuminated lamp but would be incapable of initiating an alarm because of the disabled or disconnected alarm panel. It appears that the automatic alarm system in the Doubletree fit this scenario.

It is difficult to say how long the automatic alarm system had been disabled, since no test records were available. However, it is speculated that the problem may have been present for some time. Fire department records included a complaint that was received on November 11, 1986, which reported that the central station was not manned 24 hours and that smoke detectors were not operational. The subsequent fire department inspection report dated December 4, 1986, stated that the central control station for the fire alarm was located in the PBX room, which was manned 24 hours that all smoke detectors were operating at the time of the inspection and that the alarm system was under a maintenance agreement for annual inspections (see Appendix 4). However, these inspection results are possibly subject to question.

Since the hotel changed ownership between the time of the fire department's inspection and the time of the fire, this investigation could not determine whether the alarm panel was in fact in the PBX room at the time of the fire department inspection, nor could it be determined how the fire alarm was tested by the inspector. Addressing the issue of the alarm panel location, the current building engineer reported that the automatic alarm control panel was removed from a closet on the third floor after the fire, not from the PBX room. This closet was well away from the PBX room, which was at the lobby level. Furthermore, the engineer stated that, to his knowledge, neither the panel nor the PBX room had been moved recently. Therefore, it is possible that the manual alarm annunciator panel behind the front desk was assumed by the inspector to have been connected with the automatic system.

Addressing the issue of the operability of smoke detectors, one could speculate that a fire inspector who acted on this complaint may have visually inspected smoke detectors, seen the illuminated lens, and assumed that the automatic alarm system was operational. Because of these factors, it is not unreasonable to believe that the automatic alarm may have been out of service as much as eight months prior to the fire.

In summary, one can say that one or more of the following factors may have contributed to the automatic alarm system's failure to operate properly at the time of the fire: the malfunctioning or possibly disconnected control panel located in an unoccupied area and perhaps not connected to audible devices, improper wiring, poor detector placement, and a misleading appearance of the operational status of smoke detectors.

Occupant Use Hoses

Although it was originally suspected that the security guard had attempted to fight the fire with occupant use hoses, subsequent investigation revealed that it may have been fire department personnel who removed the hoses from their racks. Given the inconsistency of the statements of the various individuals questioned, it will probably never be known for certain whether the guard actually attempted to fight the fire or not. In any case, this example opens for discussion the issue of occupant use fire extinguishing equipment.

Over the years, pressure has been mounting by many fire officials to eliminate occupant use hoses from buildings unless a fully trained and equipped fire brigade is present. Since occupant use hoses are typically inadequately inspected and tested and may encourage untrained occupants to fight fires that are beyond the capability of a small-capacity handline, they argue for limiting occupant use equipment to portable fire extinguishers. Fire extinguishers can generally control fires that are small enough for occupants to confront. Unlike occupant hoses, though, fire extinguishers encourage the user to abandon firefighting if they expire before a fire is extinguished. Codes- and standards-making bodies should begin to take a closer look at requirements for occupant hoses, given their somewhat controversial history.

Human Behavior

Convergence Clusters

A "convergence cluster" is a group of occupants who converge into a room with the intent of using it as a place of refuge until such time as they are able to escape or be rescued. The act of convergence apparently serves to reduce the anxiety and tension of group members. The concept of convergence clusters was first brought into significance by Dr. John L. Bryan of the University of Maryland. Clusters have been demonstrated to have occurred in fires such as in the Georgian Towers fire in Maryland in 1979 and in the MGM Grand Hotel fire in 1980. [1,2]

In the incident at the Doubletree Hotel, a convergence cluster appears to have occurred on the eleventh floor, based on a written statement and a news interview with an eleventh-floor occupant. As stated previously, members of the church group were in the hallway conversing when smoke began to issue from the corridor ventilation opening adjacent to Stairway 1. The group, unable to exit from the floor due to smoke and beginning to panic, apparently formed a social unit, as one individual took command and led approximately 15 people into Room 1109. According to the news report, group members calmed each other through prayer to reduce anxiety and tension until they were led to safety by hotel staff. The act of individuals deciding to remain in buildings during a fire incident rather than risking escape is significant in designing buildings for fire safety.

Lack of Belief in the Reality of Fire Alarms

Some occupants said they failed to evacuate when the fire alarm went off because of the previous series of false alarms. These occupants complained that they were not aware that there was an actual fire until they smelled smoke or were later told to evacuate.

Staff Actions

It is apparent that the security guard acted improperly by entering the tenth floor, given the conditions likely to have been present at the time. Proper action would have been to go to another floor, initiate an alarm, and assist with evacuation of guests. Also, the building engineer rode an elevator, which filled with smoke, to the seventeenth floor without any protective equipment; he could have been overcome in the elevator and should not have been using it in the fire, even though he was trying to shut off fans to help the problem.

Building Construction and Contents

As demonstrated in prior hotel fires such as the La Posada Hotel fire in McAllen, Texas,[3] the combination of light fire loading and fire-resistive construction can play a key role in limiting fire damage when automatic fire extinguishing systems are not present. In the Doubletree incident, as was true at the La Posada, the wall coverings, carpeting, and contents in the corridor did not significantly contribute to the spread of the fire. The packing materials involved in the hallway of the Doubletree were therefore somewhat isolated, and the fire was small enough so as not to pose a significant challenge to the building's fire-resistive construction.

Although the Doubletree lacked self-closing guest room doors, such closing devices would not have made much difference in this incident because the fire floor was unoccupied, and most or all of the doors were closed at the time of the fire. Performance of the doors was noteworthy: the guest rooms and storage areas on the tenth floor that had closed doors sustained only minor smoke damage in most cases (see Figures 13-15). The possible exception to door performance may have been the stairway doors, which apparently leaked significant quantities of smoke into the stairways.

The other primary means of smoke travel appeared to be the corridor ventilation system, which included air ducts adjacent to Stairway 1 and Stairway 3. Each floor had a fire-dampered (but not smoke-dampered) duct opening at either end of the corridor. The initial entry of smoke to the eleventh floor was reported to be through the duct located at Stairway 1. Fire department sources indicated that one of the two fire dampers on the tenth floor did not close completely when the fusible link operated, which would have allowed large quantities of smoke to continue spreading. Physical examination of the fire scene did not reveal additional significant means of smoke spread.

LESSONS LEARNED

The loss experienced in the Doubletree incident lends additional support to some already well-known lessons and also brings to light some emerging issues in providing fire safety to the public. These lessons address the areas of fire protection systems, building construction, fire prevention, fire department tactics, staff training, and human behavior.

Fire Protection Systems

1. **Partial Protection by Automatic Sprinklers Is Just That**

    The Doubletree Hotel was protected in many areas by an automatic sprinkler system, but the unsprinklered area provided an arsonist with the opportunity to create a major fire incident by simply lighting a match. There are many jurisdictions throughout the United States that are reducing the reliability of sprinkler systems by permitting compromises to the completeness of sprinkler protection. It should be emphasized that eliminating an area from sprinkler coverage has a direct impact on the risk of a major fire and provides an easy target to an arsonist. If corridor areas in the Doubletree had been protected as currently required by Nationally recognized standards, the magnitude of the incident would likely have been little more than an inconvenience to guests.

2. **Automatic Fire Alarm Systems Should Be Installed In A Reliable Manner And Should Be Regularly Inspected And Tested**

    The automatic fire alarm system installed in the Doubletree was not installed in compliance with the Nationally recognized standards that were in effect at the time of construction. Although there was no code requirement for the system to comply with any standards, this incident raises the issue of an owner's responsibilities when providing protection in excess of minimum code requirements. Traditionally, the model codes have held that protection in excess of the minimum requirements called for by code need not comply with any standards. Because there is likely to be a reliance on fire alarms when they are present, whether required or not, any fire alarm system should be required to be reliable.

It is incumbent upon fire and building departments to identify fire protection systems inadequacies and seek correction of these inadequacies even when such systems are installed voluntarily. There is need for fire departments to employ fore inspectors or fire protection engineers with a level of expertise in fire protection systems adequate to perform system inspections.

3. **Smoke Detector Systems Should Be Designed So That A Detector Cannot Appear Functional When The Alarm Panel Is Disabled**

The physical appearance of smoke detectors to the casual observer should be such that a detector does not display a light or signal that would lead one to believe the detector is functional when it is not.

A two-wire detector circuit that provides power and alarm-initiating capability simultaneously from a single pair of wires meets this need, as opposed to a four-wire circuit in which power is supplied by a separate power source. Fire protection professionals should consider whether the standard for smoke detection systems should be revised.

Construction Features

4. **Interior Finishes, Carpeting, And Contents In Corridors Make A Difference In Fire Safety**

By providing corridors with low flame spread finishes, ignition-resistant carpeting, and little or no combustible furnishings, the growth and intensity of a corridor fire can be limited, which provides extra escape time for occupants and keeps the fire severity manageable for arriving fire suppression forces.

5. **Subdivision of Corridors In Multi-Family Residential And High-Rise Occupancies Is Useful To Limit The Spread Of Smoke And Fire**

Smoke often travels unchecked through corridors, making escape routes impassable and allowing hazardous conditions to quickly permeate entire floors. In occupancies where people sleep or are disabled, the fire protection design that incorporates smoke doors and perhaps fire doors to subdivide corridor areas would prevent uncontrolled fire growth and smoke spread within corridors. The Doubletree Hotel did not have any such subdivisions, with the result that entire floors filled quickly with smoke from a relatively small fire and trapped a number of people.

6. **Vertical Shafts That Open To More Than One Floor Level In High-Rise Buildings, Especially Residential-Use High Rises, Should Have Smoke Dampers Installed At Each Floor Level**

Fire deaths in high-rise buildings are often caused by the inability of the building to contain smoke to a single floor level. A common path for smoke spread, which emerged again in the Doubletree incident, is a vertical ventilation duct. By installing smoke dampers at each floor level, this significant means of smoke spread would be controlled and the risk to occupants significantly reduced.

Fire Prevention and Public Education

7. Effective Hotel Staff Training Is Essential To Ensure The Safety Of Staff And Guests Alike

Although it will never be known conclusively what actions are taken by the security guard after he arrived on the tenth floor, it seems likely that he could have avoided serious injury with better training. Staff members who occupy buildings where a high risk of life or property loss exists should be fully trained in emergency procedures. Such procedures should include properly reporting a fire, initiating evacuation procedures, when to use fire extinguishers and hoses, and when to escape. In the La Posada Hotel fire, inappropriate action by the staff was also a major factor contributing to a fire's severity.[4]

8. Fire Prevention And Public Education Programs For Hotels Should Emphasize The Danger Of Storing Combustibles In Corridors

Given the many occurrences of major fires in hotels that involve combustible storage in corridors, the fire service should place additional emphasis on prohibiting this practice. Such storage poses a quick target for a would-be arsonist and, when ignited, can almost immediately block the primary egress route. Although not in a corridor, the disastrous Du Pont Plaza Hotel fire in Puerto Rico in 1986 also was started by an arsonist lighting new hotel furniture stored in its packing materials in a ballroom.

Fire Department Operations

9. High-Risk Occupancies Should Be Adequately Pre-Fire Planned

The fire scene is not the place to begin thinking of the resources and tactics necessary to handle a fire incident in a high-risk occupancy. Fire departments should identify high-risk occupancies during planning exercises and develop a course of action to be followed when an incident occurs. Such a plan should include resources necessary, resource deployment, incident command considerations such as locations for command and staging areas, and use of elevators that will be available on emergency power. In the Doubletree incident, the fire department's familiarity with the hotel aided them in controlling the incident; however, they also identified problems such as arranging for master keys to be immediately available during emergencies. Pre-arranging for resources such as keys, building plans, and owner's representatives to be available at a fire incident can be invaluable to conducting successful fire department operations.

10. Using A Supervised Rehabilitation Area Allows The Fireground Commander To Have Control Over Personnel Resources At All Times

It is becoming increasingly popular for fireground commanders to establish a rehabilitation sector at major fire incidents. The practice allows for personnel to be accounted for throughout a major incident, allows for supervision of the condition of personnel during rest periods, and provides a means to effectively monitor the availability of crews to return to service. The use of a rehabilitation sector in the New Orleans incident proved to be effective.

11. Videotaping Critiques Of Major Incidents Can Be Beneficial

Although not done for this fire, the New Orleans Fire Department indicated that it would have been beneficial to videotape the critique of this incident. Personnel who did not participate in the incident would then have a means to review the critique and gain valuable training.

CONCLUSIONS

The Doubletree Hotel fire in New Orleans will be remembered as an incident that taught us new lessons and reinforced old ones. Fire incidents in high-rise hotels should come as no surprise as long as these occupancies lack basic components of good fire protection, such as complete and functional protection by sprinklers and automatic fire alarm systems. Risk analysis and problem-solving tools such as those taught in the National Fire Academy's "Fire Risk Analysis" and "Community Fire Defenses" courses are readily available for fire service personnel to learn to identify, quantify, evaluate, and reduce the risk posed by high-risk occupancies such as high-rise hotels.

The traditional attitude about fire protection is retrospective, or "fix it after a fire occurs." This attitude must be changed to use foresight to identify existing hazardous situations and correct them. Only now has the Doubletree decided to install automatic sprinklers throughout the building.

REFERENCES

1. Bryan, John L., and Philip J. DiNenno, "An Examination and Analysis of the Dynamics of Human Behavior in the Fire Incident at the Georgian Towers on January 9, 1979." Washington, DC: Center for Fire Research, National Bureau of Standards, Department of Commerce, NBS-GCR-79-187, April 30, 1979.

2. National Fire Protection Association, "An Examination and Analysis of the Dynamics of the Human Behavior in the MGM Grand Hotel Fire." Quincy, MA, 1981.

3. U.S. Fire Administration, Federal Emergency Management Agency, "USFA Fire Investigation, La Posada Hotel Fire (McAllen, Texas)." Washington, DC, February 25, 1987.

4. Ibid.

Figure 1. The Doubletree Hotel, a 17-story high-rise.

Figure 2. PHOTOGRAPH KEY

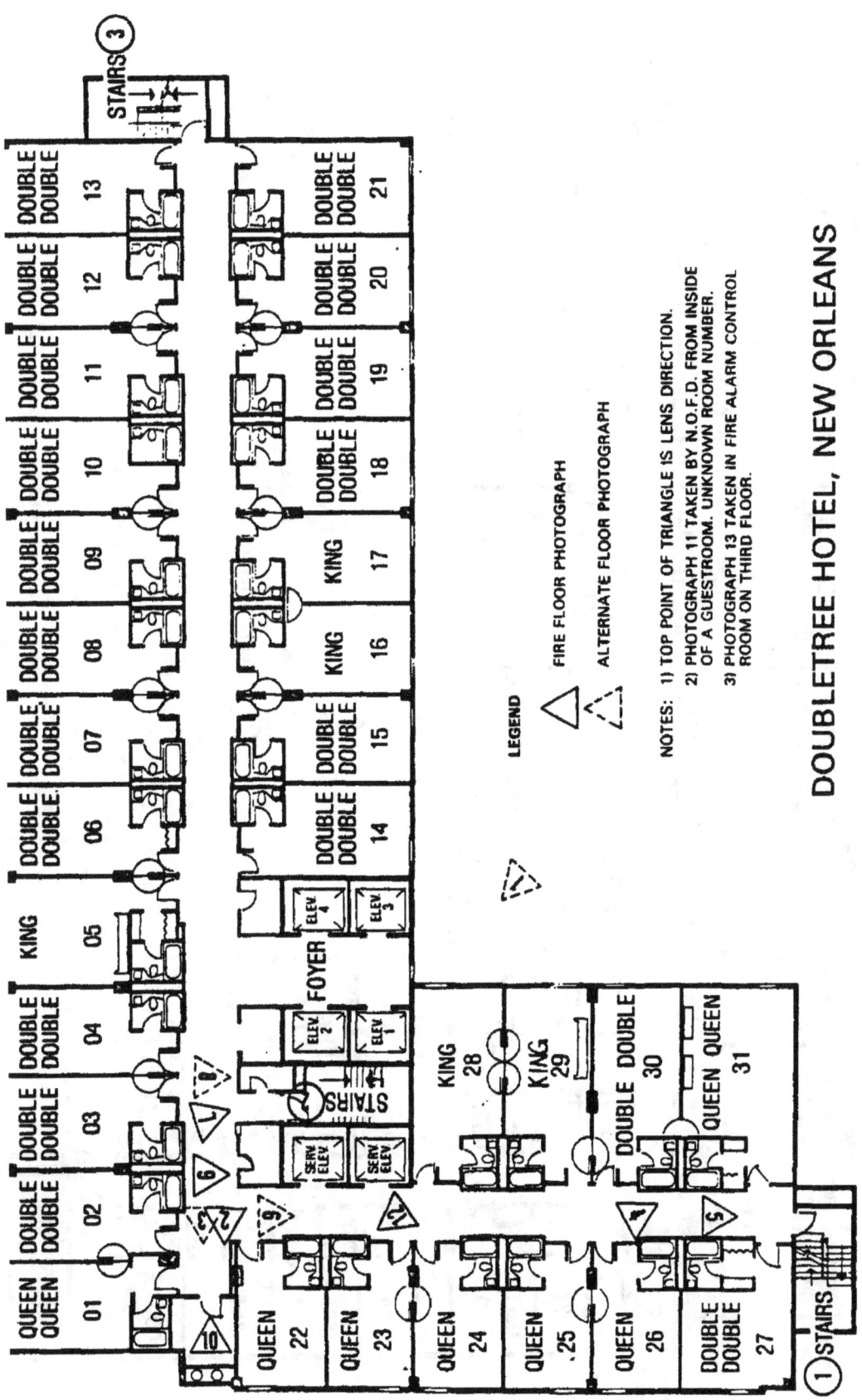

DOUBLETREE HOTEL, NEW ORLEANS

Figure 3. FLOOR PLAN: 10th FLOOR

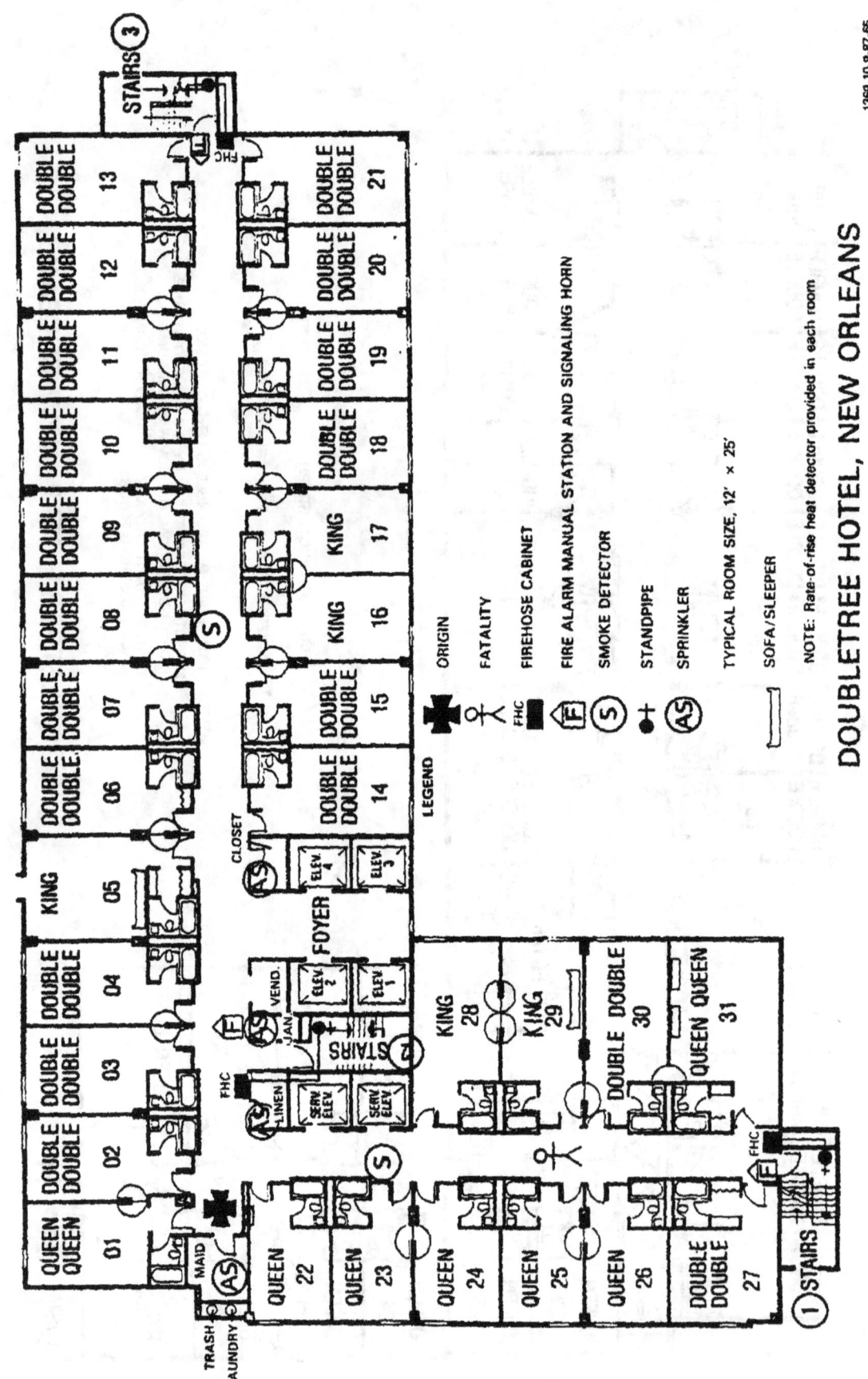

LEGEND

✠ ORIGIN

☦ FATALITY

FHC ▮ FIREHOSE CABINET

Ⓕ FIRE ALARM MANUAL STATION AND SIGNALING HORN

Ⓢ SMOKE DETECTOR

●+ STANDPIPE

ⒶⓈ SPRINKLER

▭ TYPICAL ROOM SIZE, 12' × 25'

☐ SOFA/SLEEPER

NOTE: Rate-of-rise heat detector provided in each room

DOUBLETREE HOTEL, NEW ORLEANS

1369-10 9-87-65

Figure 4. View of short corridor from area of origin toward
Stairway 1 – 9th floor.

Figure 5. Entrance to exit Stairway 1 – 10th floor. Note the ventilator
opening (upper left) and fire hose/extinguisher cabinet (lower left).

Figure 6. Entrance to exit Stairway 2 – 9th floor. Compare with Figure 7 after fire.

Figure 7. Entrance to exit Stairway 2 – 10th floor. Note pull station at left used to initiate local alarm by guard.

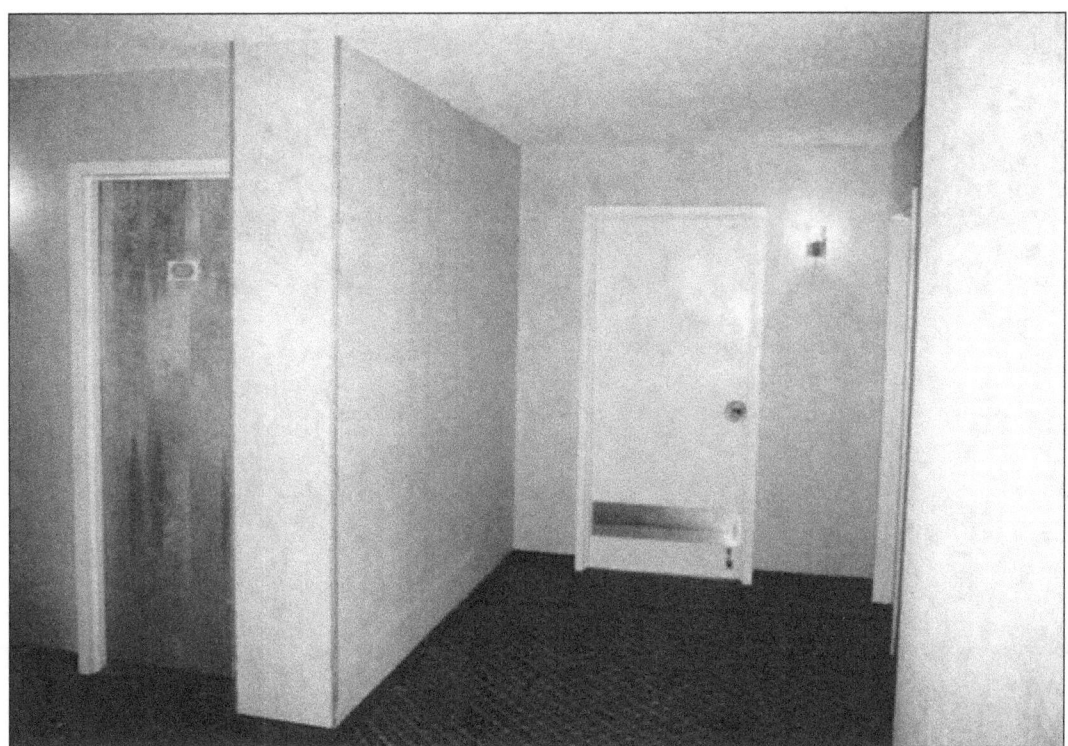

Figure 8. Area of origin on 9th floor, for comparison.

Figure 9. Area of origin on 10th floor.

Figure 10. Debris similar to that involved as first material ignited – 10th floor.

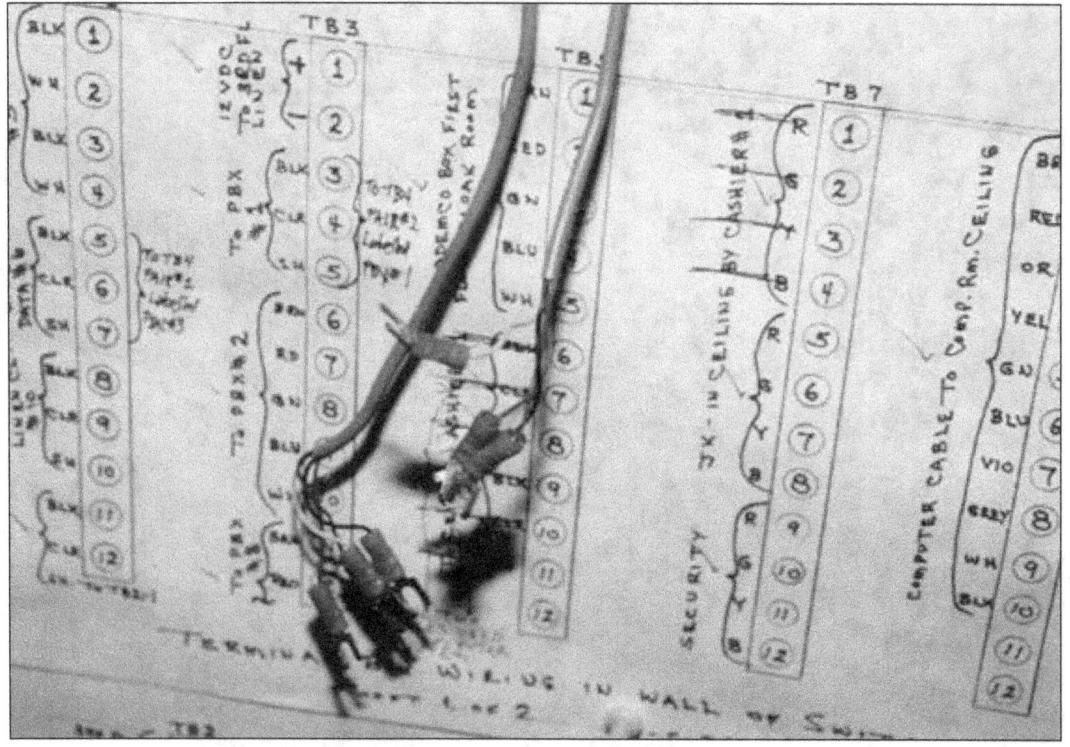

Figure 11. Telephone wire used for automatic fire alarm system, security system, and public address system.

Figure 12. Location of corridor smoke detector (dark spot on ceiling).

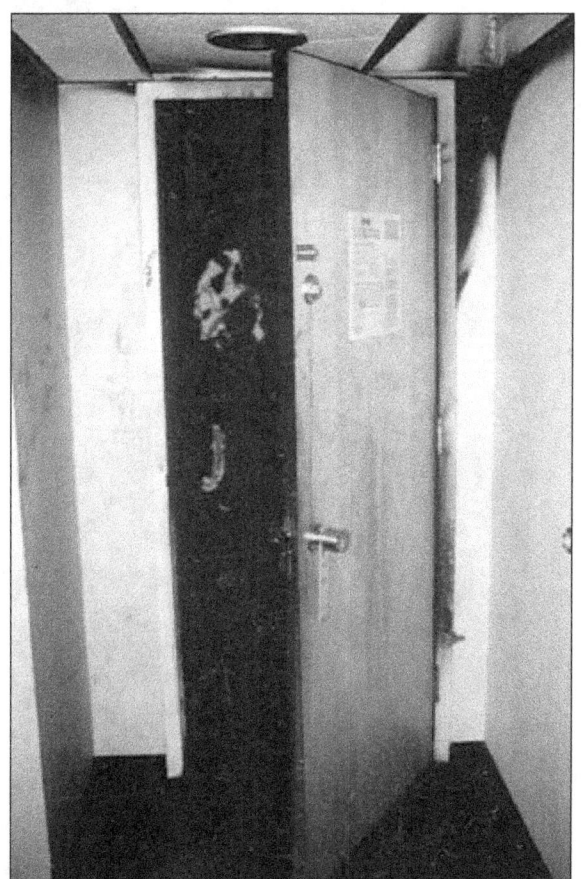

Figure 13. Guest rooms on 10th floor sustained minimal damage due to performance of doors.

Figure 14. Fire door protected linens in closet only 20 feet from area of origin.

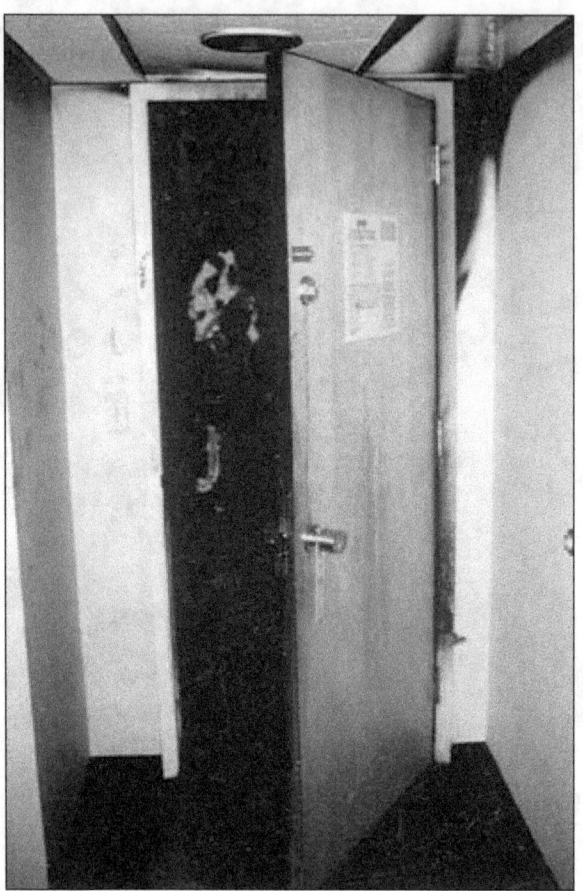

Figure 15. Sprinkler in maid's closet immediately adjacent to area of origin was not activated due to performance of fire door.

LIST OF APPENDICES

Appendix 1: FD 200 – Field Incident Report

Appendix 2: FD 500 – Casualty Report

Appendix 3: New Orleans Health Department Injury Report

Appendix 4: New Orleans Fire Department Selected Fire Prevention Records

Appendix 5: Slides of Doubletree Hotel Fire Investigation (slides with USFA master file copy only)

Appendix 6: Videotape with News Segment (file copy only)

APPENDIX 1: FD 200 – FIELD INCIDENT REPORT

DATE JULY 19, 1987   **TIME** 22:32

CHANGE 2 ☐   (73)
DELETE 3 ☐

COMPLETE FOR ALL INCIDENTS

5 0 | **INCIDENT NUMBER** 517618 | **SUPP** 1

FIELD INCIDENT REPORT

FD-200

AUXILIARY TRIP? (CHECK) | **INCIDENT TYPE** BUILDING 111 | **TYPE OF ACTION TAKEN** EXTINGUISHMENT 4 | **PROPERTY NAME** DOUBLE TREE HOTEL | **APT. NO.**

CORRECT LOCATION/ADDRESS 3101 CANAL ST. | **CITY** NEW ORLEANS | **STATE** LA. | **ZIP CODE**

COMMUNITY OR SUBDIVISION | **COUNTY** | **PARCEL NUMBER OR CENSUS TRACT** CAR 3021 PLT 3 | **DISTRICT** 2 | **OUT OF JURISDICTION (CHECK)**

COMPLEX | **FIXED PROPERTY USE** HOTEL 441

PROPERTY REPRESENTATIVE: NAME CARL McKEE | **TELEPHONE** 581-1300 | **ADDRESS (STREET, CITY, STATE, ZIP CODE)** 4124 COGNAL KENNER LA.

OCCUPANT: NAME | **TELEPHONE** | **RELATIONSHIP** | **PROPERTY MANAGEMENT** PRIVATE | CD 1

COMPLETE IF FIRE

5 1 1 | **LEVEL OF ORIGIN** TENTH STORY KAA 4110 | **AREA OF ORIGIN** HALLWAY KAB 01

EQUIPMENT INVOLVED IN IGNITION KBA | **FORM OF HEAT OF IGNITION** UNDER INVESTIGATION KBB 99

IF EQUIPMENT INVOLVED IN IGNITION | **YEAR** | **MAKE** | **MODEL** | **SERIAL NO.** | **VOLTAGE (IF ANY)**

TYPE OF MATERIAL IGNITED UNDER INVESTIGATION KCA 99 | **FORM OF MATERIAL IGNITED** UNDER INVESTIGATION KCB 99

ACT OR OMISSION — ORIGIN OF FIRE UNDER INVESTIGATION KDA 99 | **CODE VIOLATION?** 1 ☐ CHECK IF VIOLATION

MAIN AVENUES OF FIRE SPREAD UNDER INVESTIGATION LA 99

TYPE OF MATERIAL MOST RESPONSIBLE FOR FIRE SPREAD UNDER INVESTIGATION LBA 99 | **FORM OF MATERIAL MOST RESPONSIBLE FOR FIRE SPREAD** UNDER INVESTIGATION LBB 99

ACT OR OMISSION MOST RESPONSIBLE FOR FIRE SPREAD UNDER INVESTIGATION LCA 99 | **CODE VIOLATION?** 1 ☐ CHECK IF VIOLATION

COMPLETE IF MOBILE PROPERTY

5 1 4 | **MOBILE PROPERTY USE** | DC | **LICENSE NO.** | **STATE**

MAKE | **MODEL** | **YEAR** | **VEHICLE ID. NO.**

COMPLETE IF LOSS INVOLVED

5 1 5 | **ESTIMATED VALUE** | **ESTIMATED LOSS** | 5 1 6 **INSURED VALUE** | **INSURED LOSS**

STRUCTURE | (BLANKET) 18000000 | 350000 | 18000000 |

CONTENTS | | | |

INSURANCE COMPANY NAME SIMPSON WALKER, MET. LA. 831-4111

26

APPENDIX 2: FD 500 – CASUALTY REPORT

DATE	TIME	ADDRESS
7-19-87	2232	300 CANAL ST.

PAGE **OF**   **CASUALTY REPORT**   **FD-500**

MONTH	DAY	YEAR
07	19	87

(73)   CHANGE 2 ☐   DELETE 3 ☐

8 1 0	INCIDENT NUMBER	SUPP	NAME	AGE
	5 7 6 8	1	JOHN J. O'BRIEW	40

SEX MALE (M) FEMALE (F)	INJURY (I) OR DEATH (D)	CIVILIAN (C) OR FIREFIGHTER (F)	FIXED PROPERTY USE	DB	MOBILE PROPERTY USE	DB
M	I	F				

(45) CONDITION BEFORE CASUALTY	(46) ACTION CAUSING CASUALTY	(47) NATURE OF CASUALTY	(48) PART OF BODY INJURED	(49) DISPOSITION OF CASUALTY
☐ 1 Asleep	☐ 1 Caught: in, under or between; or trapped by	☐ 1 Burns and asphyxia/smoke	☐ 1 Head, Neck, includes respiratory system	☐ 1 Refused help
☐ 2 Bedridden or other physical handicap	☐ 2 Exposed to: heat, chemicals, radiation, smoke, etc.	☐ 2 Burns only	☑ 2 Body, Trunk, Back	☐ 2 First aid at scene and released
☐ 3 Impaired by drugs or alcohol	☐ 3 Fell over, on, or tripped on	☐ 3 Asphyxia/smoke only	☐ 3 Arm	☐ 3 Taken to hospital — by fire department vehicle
☐ 4 Under restraint	☐ 4 Stopped on or into	☐ 4 Wound, cut, bleeding	☐ 4 Leg	☐ 4 Taken to hospital — by non-fire department vehicle
☐ 5 Too young to act	☐ 5 Overexertion	☐ 5 Dislocation, fracture	☐ 5 Hand	☐ 5 Taken to other than hospital
☐ 6 Too old to act; senile	☑ 6 Rubbed by or contact with	☑ 6 Complaint of pain	☐ 6 Foot	☐ 6 Died
☐ 7 Mentally handicapped	☐ 7 Struck by	☐ 7 Shock	☐ 7 Internal — except respiratory system	☐ 7 Other (specify)
☑ 8 Awake and unimpaired	☐ 8 Not applicable	☐ 8 Strain, sprain	☐ 8 Multiple parts	☑ 8 Undetermined
☐ 9 Other (specify)	☐ 9 Other (specify)	☐ 9 Other (specify)	☐ 9 Other (specify)	
☐ 0 Undetermined or not applicable	☐ 0 Undetermined	☐ 0 Undetermined	☐ 0 Undetermined	

REMARKS

While hooking up supply lines to a standpipe, I had to go under a parked tractor-trailer and hit my back on a piece of iron under the trailer.

MONTH	DAY	YEAR

(73)   CHANGE 2 ☐   DELETE 3 ☐

8 1 0	INCIDENT NUMBER	SUPP	NAME	AGE

SEX MALE (M) FEMALE (F)	INJURY (I) OR DEATH (D)	CIVILIAN (C) OR FIREFIGHTER (F)	FIXED PROPERTY USE	DB	MOBILE PROPERTY USE	DB

(45) CONDITION BEFORE CASUALTY	(46) ACTION CAUSING CASUALTY	(47) NATURE OF CASUALTY	(48) PART OF BODY INJURED	(49) DISPOSITION OF CASUALTY
☐ 1 Asleep	☐ 1 Caught: in, under or between; or trapped by	☐ 1 Burns and asphyxia/smoke	☐ 1 Head, Neck, includes respiratory system	☐ 1 Refused help
☐ 2 Bedridden or other physical handicap	☐ 2 Exposed to: heat, chemicals, radiation, smoke, etc.	☐ 2 Burns only	☐ 2 Body, Trunk, Back	☐ 2 First aid at scene and released
☐ 3 Impaired by drugs or alcohol	☐ 3 Fell over, on; or tripped on	☐ 3 Asphyxia/smoke only	☐ 3 Arm	☐ 3 Taken to hospital — by fire department vehicle
☐ 4 Under restraint	☐ 4 Stopped on or into	☐ 4 Wound, cut, bleeding	☐ 4 Leg	☐ 4 Taken to hospital — by non-fire department vehicle
☐ 5 Too young to act	☐ 5 Overexertion	☐ 5 Dislocation, fracture	☐ 5 Hand	☐ 5 Taken to other than hospital
☐ 6 Too old to act; senile	☐ 6 Rubbed by or contact with	☐ 6 Complaint of pain	☐ 6 Foot	☐ 6 Died
☐ 7 Mentally handicapped	☐ 7 Struck by	☐ 7 Shock	☐ 7 Internal — except respiratory system	☐ 7 Other (specify)
☐ 8 Awake and unimpaired	☐ 8 Not applicable	☐ 8 Strain, sprain	☐ 8 Multiple parts	☐ 8 Undetermined
☐ 9 Other (specify)	☐ 9 Other (specify)	☐ 9 Other (specify)	☐ 9 Other (specify)	
☐ 0 Undetermined or not applicable	☐ 0 Undetermined	☐ 0 Undetermined	☐ 0 Undetermined	

REMARKS

NOTE: POSITIONS 50-55 OF EACH CARD MUST CONTAIN THE DATE OF THE CASUALTY.

APPENDIX 2: FD 500 – CASUALTY REPORT *continued*

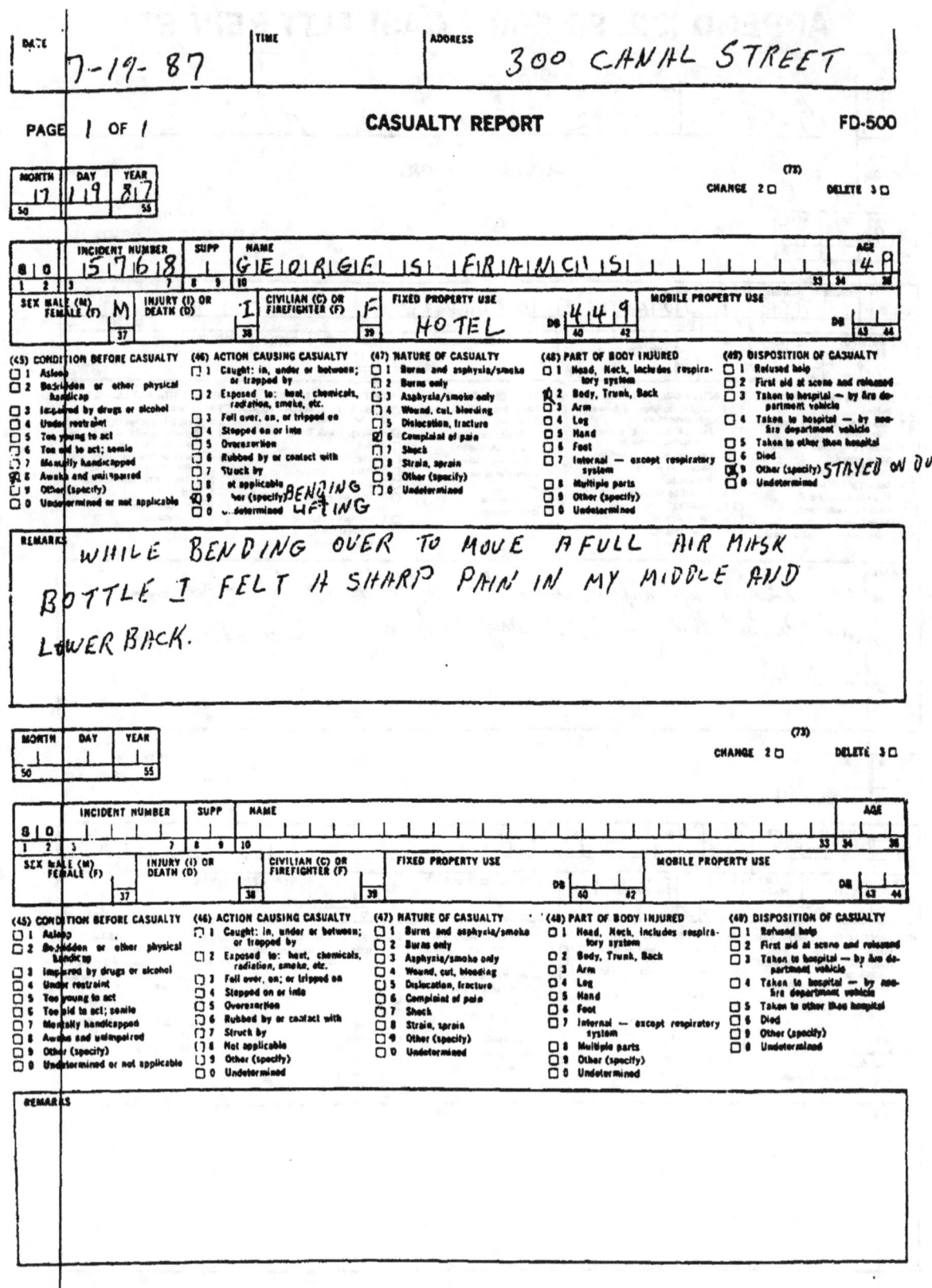

REMARKS

WHILE BENDING OVER TO MOVE A FULL AIR MASK
BOTTLE I FELT A SHARP PAIN IN MY MIDDLE AND
LOWER BACK.

NOTE: POSITIONS 50-55 OF EACH CARD MUST CONTAIN THE DATE OF THE CASUALTY.

APPENDIX 2: FD 500 – CASUALTY REPORT *continued*

DATE	TIME	ADDRESS
July 19, 1987	22:32	300 Canal St.

PAGE   OF

CASUALTY REPORT

FD-500

MONTH 7 DAY 19 YEAR 87
50 — 55

(73)
CHANGE 2 ☐   DELETE 3 ☐

8 0   INCIDENT NUMBER 57168   SUPP 1   NAME HENRY GONZALES   AGE 35
1 2 3   7 8 9 10   33 34 36

SEX MALE (M) M   INJURY (I) OR DEATH (D) D   CIVILIAN (C) OR FIREFIGHTER (F) C   FIXED PROPERTY USE HOTEL   DB 441   MOBILE PROPERTY USE   DB
        37   38   39   40 42   43 44

(45) CONDITION BEFORE CASUALTY
- ☐ 1 Asleep
- ☐ 2 Bedridden or other physical handicap
- ☐ 3 Impaired by drugs or alcohol
- ☐ 4 Under restraint
- ☐ 5 Too young to act
- ☐ 6 Too old to act; senile
- ☐ 7 Mentally handicapped
- ☒ 8 Awake and unimpaired
- ☐ 9 Other (specify)
- ☐ 0 Undetermined or not applicable

(46) ACTION CAUSING CASUALTY
- ☐ 1 Caught: in, under or between; or trapped by
- ☐ 2 Exposed to heat, chemicals, radiation, smoke, etc.
- ☐ 3 Fell over, on, or tripped on
- ☐ 4 Stepped on or into
- ☐ 5 Overexertion
- ☐ 6 Rubbed by or contact with
- ☐ 7 Struck by
- ☐ 8 Not applicable
- ☐ 9 Other (specify)
- ☒ 0 Undetermined

(47) NATURE OF CASUALTY
- ☐ 1 Burns and asphyxia/smoke
- ☐ 2 Burns only
- ☐ 3 Asphyxia/smoke only
- ☐ 4 Wound, cut, bleeding
- ☐ 5 Dislocation, fracture
- ☐ 6 Complaint of pain
- ☐ 7 Shock
- ☐ 8 Strain, sprain
- ☐ 9 Other (specify)
- ☒ 0 Undetermined

(48) PART OF BODY INJURED
- ☐ 1 Head, Neck, includes respiratory system
- ☐ 2 Body, Trunk, Back
- ☐ 3 Arm
- ☐ 4 Leg
- ☐ 5 Hand
- ☐ 6 Foot
- ☐ 7 Internal — except respiratory system
- ☐ 8 Multiple parts
- ☐ 9 Other (specify)
- ☒ 0 Undetermined

(49) DISPOSITION OF CASUALTY
- ☐ 1 Refused help
- ☐ 2 First aid at scene and released
- ☐ 3 Taken to hospital — by fire department vehicle
- ☐ 4 Taken to hospital — by non-fire department vehicle
- ☐ 5 Taken to other than hospital
- ☒ 6 Died
- ☐ 9 Other (specify)
- ☐ 0 Undetermined

REMARKS CASUALTY WAS A SECURITY GUARD

MONTH 7 DAY 19 YEAR 87
50 — 65

(73)
CHANGE 2 ☐   DELETE 3 ☐

8 0   INCIDENT NUMBER 57168   SUPP 1   NAME JOHN BURKART   AGE 52
1 2 3   7 8 9 10   33 34 36

SEX MALE (M) M   INJURY (I) OR DEATH (D) I   CIVILIAN (C) OR FIREFIGHTER (F) F   FIXED PROPERTY USE HOTEL   DB 441   MOBILE PROPERTY USE   DB
        37   38   39   40 42   43 44

(45) CONDITION BEFORE CASUALTY
- ☐ 1 Asleep
- ☐ 2 Bedridden or other physical handicap
- ☐ 3 Impaired by drugs or alcohol
- ☐ 4 Under restraint
- ☐ 5 Too young to act
- ☐ 6 Too old to act; senile
- ☐ 7 Mentally handicapped
- ☒ 8 Awake and unimpaired
- ☐ 9 Other (specify)
- ☐ 0 Undetermined or not applicable

(46) ACTION CAUSING CASUALTY
- ☐ 1 Caught: in, under or between; or trapped by
- ☐ 2 Exposed to heat, chemicals, radiation, smoke, etc.
- ☐ 3 Fell over, on, or tripped on
- ☐ 4 Stepped on or into
- ☐ 5 Overexertion
- ☐ 6 Rubbed by or contact with
- ☐ 7 Struck by
- ☐ 8 Not applicable
- ☐ 9 Other (specify)
- ☐ 0 Undetermined

(47) NATURE OF CASUALTY
- ☐ 1 Burns and asphyxia/smoke
- ☐ 2 Burns only
- ☐ 3 Asphyxia/smoke only
- ☐ 4 Wound, cut, bleeding
- ☐ 5 Dislocation, fracture
- ☐ 6 Complaint of pain
- ☐ 7 Shock
- ☐ 8 Strain, sprain
- ☒ 9 Other (specify)
- ☐ 0 Undetermined

(48) PART OF BODY INJURED
- ☐ 1 Head, Neck, includes respiratory system
- ☐ 2 Body, Trunk, Back
- ☐ 3 Arm
- ☐ 4 Leg
- ☐ 5 Hand
- ☒ 6 Foot
- ☐ 7 Internal — except respiratory system
- ☐ 8 Multiple parts
- ☒ 9 Other (specify)
- ☐ 0 Undetermined

(49) DISPOSITION OF CASUALTY
- ☐ 1 Refused help
- ☐ 2 First aid at scene and released
- ☐ 3 Taken to hospital — by fire department vehicle
- ☐ 4 Taken to hospital — by non-fire department vehicle
- ☐ 5 Taken to other than hospital
- ☐ 6 Died
- ☒ 9 Other (specify)
- ☐ 0 Undetermined

REMARKS Burning in chest + throat, no first aid given as of this writing

APPENDIX 3: NEW ORLEANS HEALTH DEPARTMENT INJURY REPORT

CITY OF NEW ORLEANS
INTER-OFFICE MEMORANDUM

Date: July 27, 1987

TO: Sheila Webb
    Deputy Director
    Department of Health

FROM: Dawne Orgeron, RN, EMT-P
      Administrator
      Health Department EMS

SUBJECT: Patient Priority Statement for Mayor's Office

As per your request I am writing our policy relative to priority of patient transports during a disaster situation such as occurred at the Double Tree Hotel on July 19, 1987.

Upon arrival at the scene the supervisor establishes a triage(sorting station) for all injured persons.

The first available ambulance unit is utilized for this purpose and does not leave the scene under any circumstances.

The second available ambulance unit and the second supervisor establish a triage station at the command post, which in this case was in the lobby of the hotel.

Persons are sorted into priority on the following basis:

First Priority: cardiac arrest, respiratory arrest, severe inhalation problems, third degree burns, second degree burns of 10% or more of their body, massive physical trauma, chest pain, stroke. IN SHORT PATIENTS DEMONSTRATING APPARENT OR IMPENDING LIFE THREAT.

Second Priority: moderate respiratory distress, second degree burns of less than 10% of their body, moderate physical trauma (no symptoms of shock), inhalation problems. PATIENTS DEMONSTRATING POTENTIAL FOR COMPROMISE OR POSSIBLE LIFE THREAT IS DELAYED AN ORDINATE PERIOD OF TIME.

Third Priority: mild respiratory distress, first degree burns, minimal trauma. PATIENTS WHO ARE CONSIDERED WALKING WOUNDED.

In any disaster situation where there are numerous deaths or fatalities, these persons are not taken off the scene and are removed from the priority one category. A

APPENDIX 3: FD 500 – INJURY REPORT *continued*

Page 2

temporary morgue is instituted at the scene to accommodate these persons.

In the situation at the Double Tree the following occurred:

1. An external triage station was established by a supervisor and Unit 6205. They managed all priority two and three patients after they were evaluated by the internal triage team.

2. An internal triage station was established by a supervisor and Unit 6201 in the lobby of the hotel, which was the fire command post.

3. Unit 6203 was on standby for transport of patients on site.

4. West Jefferson Ambulance Service was asked to cover the West Bank and Unit 6204 was pulled to centralize at Charity Hospital New Orleans.

5. St. Bernard Parish Ambulance Service was asked to cover New Orleans East and Unit 6207 was pulled to centralize at Charity Hospital New Orleans.

6. Emergency One Ambulance from St. Tammany Parish was asked to cover Irish Bayou and the Twin Span Bridge.

7. Medic One Ambulance was asked to cover all Code 1 traffic which occurred.

8. Two calls were rolled to Medic One, both of which were cancelled by NOFD prior to their arrival. No calls were handled by the other mutual aide services.

9. The first patient transferred off the scene was the security guard who was in cardiac arrest. He was taken to Tulane Medical Center.

10. The second transport off the scene was with two patients both of whom claimed smoke inhalation. They were treated with oxygen on the scene, both had normal respiratory rates, and good vital signs.

11. The third transport off the scene was with a young lady who presented at 0040 hours stating she burned her hand on a curling iron. She was transported immediately.

APPENDIX 3: FD 500 – INJURY REPORT *continued*

Page 3

12. The following patients were treated prior to the second unit leaving and refusals were obtained from all:

    1 Double Tree Employee
    2 New Orleans Police Officers
    3 minors who had guardian consent for refusal.
    Total of 6 patients.

The call was received at 2240 hours.

The first unit left the scene with the cardiac arrest at 2335 hours

The second unit left the scene with the two minors at 0003 hours.

The third unit left the scene with the minor with the burned hand at 0045 hours.

The only critical patient requiring immediate transport to a medical facility which presented to our personnel was the man in cardiac arrest. All other patients evaluated were minor, and non life threatening. THe majority of the incidents which occurred at this scene were emotional because of the situation. We remained on the scene until 0100 hours at which time the fire department dismantled their command post.

If you have any further questions relative to this matter, please do not hesitate to contact me.

APPENDIX 3: FD 500 – INJURY REPORT *continued*

Page 4

Breakdown of Patient Reports at Double Tree Hotel
Sunday July 19, 1987

1. Male age 35, Dx. Cardiac, Respiratory Arrest.
   discovered in building by fire department at 2305, EMS
personnel went up to patient arrived at 2308, brought down
to unit at 2335. Arrived Tulane Medical Center at 2340
hours. Five personnel involved in this patient's care.
Transported Code 3.

2. Male age 41, Dx. Requesting Oxygen
   presented at 2400 hours, ambulatory requesting oxygen
after slight inhalation of smoke. Is a police officer. VS
144/88, HR 92, RR 24 **REFUSED TRANSPORT**.

3. Male age 32, Dx. Requesting Oxygen
   presented at 2400 hours, ambulatory requesting oxygen
after slight inhalation of smoke. Is a police officer. VS
136/78, HR 88, RR 24 **REFUSED TRANSPORT**.

4. Male age 57 Dx. Possible Smoke Inhalation.
   presented at approximately 2300 c/o of smoke
inhalation. No trauma noted, no soot in airway or nares.
Employee of Double Tree Hotel. VS 140/80, HR 110, RR 20.
Administered oxygen, 2308 second VS 136/80, HR 100, RR
20. **REFUSED TRANSPORT**

5. Female age 14 Dx. Possible Smoke Inhalation
   presented at approximately 2315 c/o smoke inhalation.
VS 110/80, HR 80, RR 24. Administered oxygen. 2320 VS
110/80, HR 80, RR 20; 2329 110/80, HR 80, RR 20. **REFUSED
TRANSPORT** signed by guardian.

6. Male age 12 Dx. Possible Smoke Inhalation
   presented at approximately 2330 c/o smoke inhalation.
No trauma noted. VS 120/90, HR 84, RR 20. Administered
oxygen. 2340 VS 122/90, HR 80, RR 20. **REFUSED TRANSPORT**
signed by guardian.

APPENDIX 3: FD 500 – INJURY REPORT *continued*

Page 5

7. Female age 14 Dx. Hyperventilation
   presented at approximately 2245 c/o hyperventilation X
5 minutes, denies smoke inhalation. No trauma noted.
Given brown paper bag to breathe in. VS 122/80, HR 100, RR
34. 2300 VS 120/80, HR 92, RR 30; 2315 VS 118/78, HR 92,
RR 20. **REFUSED TRANSPORT signed by guardian.**

8. Female age 12 Dx Possible Smoke Inhalation
   presented at 2330 c/o smoke inhalation. VS 100/p, HR
130, RR 20, administered oxygen. 2345 VS 100/p, HR 126, RR
20; 2357 VS 100/p, HR 126, RR 20. **Transported Code 1 to**
Tulane Medical Center by Unit 6203. Left scene at 2359,
arrived at 0003.

9. Female age 14 Dx. Possible Smoke Inhalation
   presented at 2340 c/o general weakness and feeling
nervous. Stated inhaled some smoke, administered oxygen.
VS Refused BP, HR 80, RR 24 Air entry good bilaterally no
wheezes. 2348 VS still refuses BP, HR 82, RR 24; 2354 VS
still refuses BP, HR 80, RR 24. **Transported Code 1 to**
Tulane Medical Center by Unit 6203. Left scene at 2359,
arrived 0003.

10. Female age 16 Dx. Pain Right Hand
    presented at approximately 0040 hours, stated she
burned her hand on curling iron while exiting Double Tree
Hotel during fire. Denies smoke inhalation, lungs clear
bilaterally. VS 112/84, HR 76, RR 21. **Transported Code 1**
to Tulane Medical Center by Unit 6205. Left scene at
0045, arrived at 0048 hours.

APPENDIX 4: NEW ORLEANS FIRE DEPARTMENT
SELECTED FIRE PREVENTION RECORDS

FII ' PREVENTION DIVISI N
N.O.F.D.

#0619

DATE ___11/11/86___

TO: _____
Inspector

Please accept the following assignment:

INTERNATIONAL HOTEL

ADDRESS ___300 CANAL ST. — 581-1300___

REMARKS: _____

1) CENTRAL CONTROL STATION NOT MANNED 24 HRS.
) SMOKE DETECTORS NOT IN OPERATION.
3) ENTRANCE DOORS (ST. PETER & TCHOUPITACOUS) BARRICADED.
) TELEPHONE OPERATORS NOT FAMILIAR WITH REPORTING FIRES.

Complainant: Benjamin Bullot

APPENDIX 4: FIRE PREVENTION RECORDS *continued*

Fire Prevention Di ion   *Page 1 of 2*   re District   *2 r a*

NEW ORLEANS FIRE DEPARTMENT
317 Decatur Street   *FILE*   Company   *F.P.O*
Telephone No. 581-5457   Time of
NEW ORLEANS. LOUISIANA   Inspection   *2:45 pm*

INSPECTION AND/OR INVESTIGATION REPORT

Date   *12/4/86*   *N.O. La. 70130*

 rty Owner's Name *Roy A. Frizzo*   Address *300 Canal St.*   Phone *581-1300*

   at *International Hotel*   Address *Same*   Phone *Same*

 er of Stories *16*   Number of Rooms *Various* Number of Persons Accommodated *Various*

 d Building Construction *1*   Type of Occupancy *Residential*

BRIEFLY EACH VIOLATION AND/OR HAZARDOUS CONDITION NOTED WHICH IN YOUR OPINION
 AUSE A FIRE OR RESULT IN INJURY TO SOMEONE SHOULD A FIRE OCCUR THERE.

@ Bill Robert, Director of Security

In response to items on complaint #0619

*1) Central control station for alarm system is
located in the PBX room so therefore it is monitored
on a 24 hr. basis.*

*2) All smoke detectors are operating at the
present time. Hotel has a maintenance agreement
with local security co. to check all detectors in
building at least once a year.*

*3) Entrance door in lobby on Tchoupitoulas
st. side of building is braced in closed position to keep
vagrants from entering building, but can be easily
opened from inside if exiting is necessary.*

*4) All telephone operators with hotel are
briefed on reporting emergencies on PBX equipment*

For File only   *John McLemore*

                                        Inspection and/or Report by

Consult reverse side of form for instructions.

APPENDIX 4: FIRE PREVENTION RECORDS *continued*

Fire Prevention Division   *Page 2 of 2*

NEW ORLEANS FIRE DEPARTMENT
317 Decatur Street
Telephone No. 581-5457
NEW ORLEANS. LOUISIANA

Fire District _____

Company _____

Time of
Inspection _____

INSPECTION AND/OR INVESTIGATION REPORT

Date _____

Property Owner's Name _____ Address _____ Phone _____

Occupant *International Hotel* Address *300 Canal St.* Phone *581-1300*

Number of Stories _____ Number of Rooms _____ Number of Persons Accommodated _____

Type of Building Construction _____ Type of Occupancy _____

DESCRIBE BRIEFLY EACH VIOLATION AND/OR HAZARDOUS CONDITION NOTED WHICH IN YOUR OPINION
MAY CAUSE A FIRE OR RESULT IN INJURY TO SOMEONE SHOULD A FIRE OCCUR THERE.

@ Bill Roberts, Director of Security

At time I responded to complaints I felt that they were unfounded.

Partial inspection

Approx 12,000 sq. ft.

No fee.

For File only

Complaint slip # 0619 attached

John McLemore

Inspection and/or Report by

APPENDIX 5: SLIDES OF DOUBLETREE HOTEL FIRE INVESTIGATION NEW ORLEANS

Slide 1   The area of origin. The room behind the area of origin is a small utility and maid's closet. The boxes that were ignited were stacked on the right side looking into the slide. The floor was unoccupied at the time of the fire and was undergoing renovation. The cardboard boxes with some foam padding were flattened and stacked against the wall. About eighteen to twenty cartons were there, in a space about 4 feet by 6 feet.

Slide 2   Area of origin looking down the long corridor. The smoke detector in the picture has been added since the time of the fire.

Slide 3   Inside of a utility closet immediately adjacent to the area of origin. The smoke and fire did not fully penetrate into the room. This shot also highlights that the floor penetrations were very well sealed, which would help prevent the passage of smoke from floor to floor.

Slide 4   The interior of the door jamb on the same utility closet as in the last slide. There was some heat and smoke damage above the door, but it was extremely limited.

Slide 5   Similar to 4.

Slide 6   A cross-section of the door jamb leading to the utility closet in the previous slide. Shows the heavy charring on the outer side and how the inner side (behind the door) is relatively less damaged.

Slide 7   The inside of Room 1001, which is adjacent to the area of origin. Again, you can see that the heat and smoke damage is very limited around the top of the door.

Slide 8   A closer shot of the same room highlights the smoke and heat damage around the top of the door.

Slide 9   This shot is the elevator lobby in the short corridor that is immediately adjacent to the area of origin.

Slide 10   This shows the wall construction of the typical corridor wall, being 2-by-4 metal studs that were approximately 24 inches on center. The drywall was 5/8 drywall, but we were unable to determine whether it was Type X, as required for fire-resistant assemblies. There was no marking indicating that.

Slide 11   This is another shot of the same elevator lobby in the short corridor, showing the heat damage to the elevator doors. Obviously, the drywall has been stripped, but the elevator doors give an indication as to how bad the damage was.

Slide 12   The door remaining on Room 1026 at the time of the investigation. At the time of the investigation, the doors had been removed from most of the rooms on the tenth floor except Room 1026. As you can see, the damage to the door here was relatively limited, as the door was well away from the area of origin. Later investigations determined that the doors were 1-3/4-inch solid-core wood.

Slide 13   Side view of the same door showing the minor level of damage of the door going through the jamb.

Slide 14   Typical HVAC draws return air from within the room foyer and exhausts supply air into the room through the grill.

Slide 15   This is a rate-of-rise heat detector that is typical for all of the guest rooms. The location is not preferable because it is in an area that is close to a dead air space, in a corner, but it does meet code.

Slide 16   This is a shot of the end of the short corridor adjacent to the area of origin. Note that the hose was partially pulled off at the time of the fire and completely pulled off later on. Above the hose cabinet is an air-circulating duct for the corridor, which was significant in the smoke spread from floor to floor. Again, notice that the damper there is closed. The fusible link had fired on that particular damper and closed it. However, the smoke still penetrated. To the side of the door is a manual pull station.

Slide 17   This is a closer shot of the hose cabinet and the air damper.

Slide 18   This is a shot of the stairwell entry area on the center stairwell closest to the area of origin. Again, you can see the level of heat damage on the door. To the left of the entry foyer to the stairway door is a fire hose cabinet, which had 1-1/2-inch hose inside. To the other side is the manual pull station that was used to originally alarm the occupants.

Slide 19   Same as 18, but with a different exposure.

Slide 20   Close-up of the manual pull station that was first used to alert the occupants from the center stairwell.

Slide 21   The area immediately adjacent to the center stairwell. It is a storage room that was sprinklered.

Slide 22   The main passenger elevator lobby, which is somewhat remote from the area of origin. The level of damage here was not great.

Slide 23   One of the storerooms on the tenth floor. You can see the heat damage above the door. However, the sprinkler did not fuse. The sprinkler system in the building was limited to small storage rooms, maids' closets, and the public assembly areas and the kitchen on the upper levels.

Slide 24   The doorway entering the room adjacent to the center stairway.

Slide 25   Another shot of the same jamb.

Slide 26   The center stairwell door, shot at a reverse angle showing the storage closet on the side.

Slide 27   A shot down the long corridor taken from the location of the center stairway in the back across the elevator lobby.

Slide 28   One of the other doors that have remained on the floor after the attorney had subpoenaed all the doors. Again, you can see that this was remote from the area of origin, and the level of damage on this particular door is somewhat limited.

Slide 29   This is a shot taken in the commercial kitchen on the sixteenth floor of the building, which is the highest occupied floor under the mechanical penthouse. Notice the sprinkler system and the range hood system.

Slide 30   The public assembly area located on the sixteenth floor, which was unoccupied at the time of the fire. This area was sprinklered. However, had this area been occupied at the time of the fire, the occupants might have been in some danger because they were above the fire and there was smoke in all three stairwells of the building.

Slide 31   The inside of the stairwell from the seventeenth floor, looking back to the sixteenth floor. It shows the standpipe system and that the construction of the stairway is noncombustible masonry, with concrete walls. On the back wall is the sprinkler valve assembly for the sixteenth floor.

Slide 32   Heat detector located in the stairway. It was part of the automatic system, which was nonfunctional at the time of the fire.

Slide 33   The ventilation opening at the top of the stairway. One was located in each stairway. It did not have a power fan on it; it just allowed any air that gravitated into the stairway to be exhausted to the outside. It would also have allowed smoke to exit.

Slide 34   The seventeenth-floor mechanical penthouse. The main fan controls were upstairs. Fortunately, at the time of the fire, the person on-duty was able to shut down the fans when directed to do so by the fire department. However, he was trapped for a short while on this floor and was unable to get out from the stairway due to heavy smoke. At the instruction of the maintenance manager, who was on the ground level, he was able to go through a penthouse area to another stairway and escape.

Slide 35   The stain left by the location of the smoke detector that was located in an alcove about midway down each corridor, the alcove being the area where four guest rooms open onto the corridor. As you can see, the smoke detector was located in a dead air space that does not meet code because it is too close to both walls in the corner.

Slide 36   Another shot of the same stain.

Slide 37   A typical guest room floor showing the condition before the fire. This was shot looking down the short corridor from the area of origin.

Slide 38   This is again on a typical guest room floor showing the original appearance of the center stairway entrance with the exit sign, the pull station, the horn above the pull station, the fire hose cabinet on the right wall, and the fire extinguisher cabinet on the left wall.

Slide 39   This shows how the area of origin would have appeared before the fire. The doorway straight ahead was the doorway into the maids' closet.

Slide 40   This shows how the floor of origin would have appeared looking down the short corridor from the area of origin. The service elevators are partway down the hall.

Slide 41   The carpeting and wall covering used in the corridors, immediately adjacent to the service elevator. There is a relatively low-nap less-combustible carpet, and the wall covering is extremely thin glued directly to noncombustible drywall. The carpeting and wall covering performed very well in limiting the extent of fire damage in the corridor.

Slide 42 Close-up of typical wall covering used in the corridor.

Slide 43 The fire alarm control panel for the manual Simplex system, which was installed about 1973. It was a high voltage non-power limited system that used all relays. It was not solid-state. The system did function at the time of the fire and was credited with alerting many of the occupants.

Slide 44 The area where, according to the maintenance man, the automatic fire alarm control panel had been located. Both of the fire alarm panels were located in the storage closet on the third floor. To the best of the investigator's determination, the automatic alarm panel was not connected to any audible alarm, nor was it monitored in any location other than this closet. Subsequent testing by the fire marshal and the Simplex Fire Alarm Company the day after the fire included blowing smoke into smoke detectors on the floors, but was unable to cause the alarm panel to go into alarm.

Slide 45 Close-up of the fire alarm wiring. According to the visual examination of the detectors, the wiring consisted of a single-conductor telephone wire that would not meet code. The panel did appear to have correct connections.

Slide 46 According to the maintenance man, this was one of the modules that were part of the fire alarm system. It appeared to be part of a public address system. A final determination could not be made at the scene.

Slide 47 The mechanical room containing the fire pump for the building. The fire pump supplies the occupant-use hose cabinets and the fire department standpipe.

Slide 48 The front side of the building, main entrance, giving a perspective of the total building. The lower portion is wider than the tower containing the lobby and reception areas.

Slide 49 The type of material involved in the ignition. You can see the cardboard box and the padding that was used to pack the furniture. These particular boxes contained wardrobes or some large piece of furniture that was being placed in each room during the renovation.

Slide 50 One of the guest room doors taken from the inside by the fire marshal. It appears to be in an area not too distant from the area of origin. As you can see, the leakage around the door was minimal. Damage to the front and the jamb of the door was significant. However, the door did hold during the fire.

Slide 51 The linen storage closet. Note that there was no real heat damage inside the closet at all. The towels are still intact and clean. The door, which was metal, appeared to be a 1-1/2-hour rating. Even though it was substantially damaged on the outside, it did not allow the fire to pass into the closet.